YOUR KNOWLEDGE HAS VALUE

- We will publish your bachelor's and master's thesis, essays and papers

- Your own eBook and book -
  sold worldwide in all relevant shops

- Earn money with each sale

Upload your text at www.GRIN.com
and publish for free

Imprint:

Copyright © 2014 GRIN Verlag, Open Publishing GmbH
Print and binding: Books on Demand GmbH, Norderstedt Germany
ISBN: 978-3-668-17883-0

This book at GRIN:

http://www.grin.com/en/e-book/276575/price-elasticity-of-demand-and-its-effect-
on-revenue

Stefanie Mücka

Price Elasticity of Demand and its effect on Revenue

GRIN Publishing

An Investigation into Price Elasticity of Demand and its effect on Revenue

Stefanie Mücka
Mathematics Higher Lever
12 MS

Rationale

I have chosen to focus my mathematical exploration on applications of Calculus in Business situations. To begin with, I was looking for an interesting real life situation I could base my investigation on. Having lived in an economically well developed country like Germany for almost my whole life, the accessibility to a wide range of products and their varying appeal to the consumer are subconsciously part of my daily life. The fact that some products are enormously demanded by society whereas others aren't that successful on the market gave me the idea to investigate how demand is influenced by outside factors. Thereby I discovered the concept of Price Elasticity of Demand which is useful in indicating the responsiveness of the demand of a certain good to a change in its price. I thus decided to explore the different levels of Price Elasticity of demand, namely elastic, inelastic and unit elasticity, and their effect on revenue by means of both an exponential and a quadratic demand function. Finally I applied the acquired knowledge to a highly demanded and very popular product in Germany, which is coffee, and modeled its change in demand dependent on varying prices as well as outside factors such as brand loyalty and income.

Introduction

Price fluctuation is a phenomenon we frequently encounter in our daily life. Therefore, Price Elasticity of demand is a useful indicator of the relation between a change in price and the willingness to purchase a certain product. In general it is proposed that an increase in price of a certain product results in a decrease in demand, however this concept is also dependent on several factors such as the availability of substitute products, the indispensability of products or brand loyalty.
Moreover, Price Elasticity of demand provides information about the revenue of the product sold at a certain price, as revenue and Price Elasticity are calculated with the same variables, namely the price p and the quantity demanded q. In regard of the elasticity of the product, an increase in price can entail both an increase and decrease in revenue, differing always in relation to its demand. Consequently, it is of immense importance to companies to model demand functions in order to foretell the implications of a change in price on demand and revenue.

The aim of this exploration is to look at the concept of Price Elasticity of demand and its effect on revenue in order to determine whether a product's price is elastic, inelastic or of unit elasticity. Thereby I will explore different types of demand functions and model their optimum price as well as maximum revenue. This concept is eventually applied to a real life situation.

To begin with, the responsiveness of demand to price change can be measured as a ratio of the percentage rate of change in the quantity that is demanded to the percentage rate of change in the price of the good. [1]

---

[1] http://www.ncssm.edu/courses/math/apcalcprojects/econ/
Elasticity_of_Demand_Student_Handout.pdf

Defining the quantity demanded as q and the price for the good demanded as p, and relying on the equation that the rate of change of a quantity

$$Q(x) = \frac{100 Q'(x)}{Q(x)} \quad \text{[2]}$$

the rate of change of $\quad q = \dfrac{100 \Delta q}{q} = \dfrac{\left(\dfrac{100 dq}{dp}\right)}{q} \quad$ and

the rate of change of $\quad p = \dfrac{100 \Delta p}{p} = \dfrac{\left(\dfrac{100 dp}{dp}\right)}{p} = \dfrac{100}{p}$

Hence, Price Elasticity of Demand

$$E(p) = \frac{\dfrac{\dfrac{100 dq}{dp}}{q}}{\dfrac{100}{q}} = \frac{p}{q} \frac{dq}{dp}$$

As it can be assumed that an increase in price will lead to a decrease in demand and that q > 0 and p > 0, E(p) has to be negative, thus $\dfrac{dq}{dp} < 0$.

As mentioned before, the Price Elasticity is largely dependent on the good, therefore there are 3 different levels of demand.[3]

- In case that $|E(p)| > 1$, the demand is said to be elastic, indicating that the percentage increase in price is less than the resulting percentage decrease in demand, hence demand is sensitive to price change
- In case that $|E(p)| < 1$, the demand is said to be inelastic, indicating that the percentage increase in price is greater than the resulting percentage decrease in demand, hence demand is rather insensitive to price change
- In case that $|E(p)| = 1$, the demand is said to be of unit elasticity, indicating that percentage change in price and demand are relatively equal [4]

In order to connect E(p) and R(p) in one equation, the rate of change of revenue with respect to price has to be found by differentiating R(p)=p×q implicitly with respect to p while q is a constant, using the product rule.

$$\frac{dR}{dp} = \frac{d}{dp}(pq) \quad = p \frac{dq}{dp} + q$$

---

[2] https://highered.mcgraw-hill.com/sites/dl/free/0073051918/297329/ch02.pdf, p.117

[3] http://www.ncssm.edu/courses/math/apcalcprojects/econ/Elasticity_of_Demand_Student_Handout.pdf

[4] http://www.economicshelp.org/blog/7019/economics/examples-of-elasticity/

Subsequently, $E(p)=\frac{p}{q}\frac{dq}{dp}$ is substituted into $\frac{dR}{dp}$ by multiplying by $\frac{q}{q}$

$$\frac{dR}{dp}=\frac{q}{q}\left(p\frac{dq}{dp}+q\right)=\frac{pq}{q}\frac{dq}{dp}+\frac{q^2}{q}$$

$$\frac{dR}{dp}=q\left(\frac{p}{q}\frac{dq}{dp}+1\right)$$

$$\frac{dR}{dp}=q\left[E(p)+1\right]$$

•In case that $|E(p)|>1$, $E(p)<-1$ as $E(p)<0$, hence $|E(p)+1|<0$

=> q $|E(p)+1|<0$ => $\frac{dR}{dp}<0$

=>demand is elastic and increase in p results in decrease of R

•In case that $|E(p)|<1$, $-1<E(p)<0$,

=> $|E(p)+1|>0$ => $\frac{dR}{dp}>0$

=>demand is inelastic and increase in p will result in increase of R

•In case that $|E(p)|=1$, $E(P)=-1$

=> $\frac{dR}{dp}=0$

=>demand is of unit elasticity and increase in p results in an relatively unchanged R

By way of example, an exponential function can express demand as a function of price.

Hence, $D(p)=q=Ae^{-kp}$, where A and k are constants.

$E(p)=\frac{p}{q}\frac{dq}{dp}$, so $\frac{dq}{dp}=-Ake^{-kp}$

$$E(p)=\frac{-Akpe^{-kp}}{Ae^{-kp}}\qquad =k\times p$$

For E(p) to be elastic, $|E(p)|>1$ and as E(p) has to be negative, $E(p)>-1$

$$\frac{p}{q}\frac{dq}{dp}>-1$$

$$-\frac{p}{q}\frac{dq}{dp}>1$$

$$\frac{dq}{dp}<-\frac{q}{p}$$

$$\frac{dq}{dp}+\frac{q}{p}<0$$

$$p\frac{dq}{dp}+q<0$$

$$\frac{dR}{dp} < 0$$, what is indicating that p > optimum price

For E(p) to be inelastic, $|E(p)| < 1$, so $-1 < E(p) < 0$

$$\frac{p}{q}\frac{dq}{dp} < -1$$

$$-\frac{p}{q}\frac{dq}{dp} < 1$$

$$\frac{dq}{dp} > -\frac{q}{p}$$

$$\frac{dq}{dp} + \frac{q}{p} > 0$$

$$p\frac{dq}{dp} + q > 0$$

$$\frac{dR}{dp} > 0$$, what is indicating that p < optimum price

For E(p) to be of unit elasticity, $|E(p)| = 1$, $E(p) = -1$

$$\frac{p}{q}\frac{dq}{dp} = -1$$

$$-\frac{p}{q}\frac{dq}{dp} = 1$$

$$\frac{dq}{dp} = -\frac{q}{p}$$

$$\frac{dq}{dp} + \frac{q}{p} = 0$$

$$p\frac{dq}{dp} + q = 0$$

$$\frac{dR}{dp} = 0$$, what is indicating that p = optimum price

Regarding the consequences on the revenue,

$$\frac{dR}{dp} = p\frac{dq}{dp} + q$$

$$= -Akpe^{-kp} + Ae^{-kp} \qquad = Ae^{-kp}(1-p)$$

Maximum revenue would occur when $\frac{dR}{dp} = 0$

$$Ae^{-kp} - Akpe^{-kp} = 0$$
$$Ae^{-kp} = Akpe^{-kp}$$
$$\ln e^{-kp} = \ln kpe^{-kp}$$
$$\ln e^{-kp} = \ln kp + \ln e^{-kp}$$
$$-kp = \ln kp - kp$$
$$0 = \ln kp \qquad\qquad \therefore k \times p = 1 \quad as \ \ln 1 = 0$$

=> In case there is an increase in p, R(p) will be left relatively unchanged

Increase in revenue would occur at inelastic demand when $\frac{dR}{dp} > 0$

$$Ae^{-kp} - Akpe^{-kp} > 0$$
$$\therefore \ k \times p > 1$$

=> In case there is an increase in p, R(p) will also increase

Decrease in revenue would occur at elastic demand when $\frac{dR}{dp} < 0$

$$Ae^{-kp} - Akpe^{-kp} < 0$$
$$\therefore \ k \times p < 1$$

=> In case there is an increase in p, R(p) will decrease

In order to find R(p), $\frac{dR}{dp}$ has to be integrated

$$\frac{dR}{dp} = -Akpe^{-kp} + Ae^{-kp}$$
$$R(p) = \int \frac{dR}{dp}\, dp \ = \ \int \left(-Akpe^{-kp} + Ae^{-kp}\right) dp$$
$$= -Ak \int \left(pe^{-kp}\right) dp + A \int \left(e^{-kp}\right) dp$$

Looking at the first Integral $\int \left(pe^{-kp}\right) dp$ of the sum of Integrals, it has to be integrated using Integration by parts.

Applying the Integration by parts formula $\int uv'\, dx = uv - \int u'v\, dx$, in this case

$u = p$, hence $u' = 1$ and $v' = e^{-kp}$, hence $v = -\frac{1}{k}e^{-kp}$

Therefore,

$$\int \left(pe^{-kp}\right) dp = p\left(-\tfrac{1}{k}e^{-kp}\right) - \int \left(-\tfrac{1}{k}e^{-kp}\right) dp$$
$$= -\tfrac{p}{k}e^{-kp} + \tfrac{1}{k}\int \left(e^{-kp}\right) dp$$
$$= -\tfrac{p}{k}e^{-kp} + \tfrac{1}{k}\left[-\tfrac{1}{k}e^{-kp}\right] + c$$
$$= -\tfrac{p}{k}e^{-kp} - \tfrac{1}{k^2}e^{-kp} + c$$

Integrating the second Integral,

$$\int \left(e^{-kp}\right) dp = \left[-\tfrac{1}{k}e^{-kp}\right] + c$$

Then the two Integrals have to be put again into one equation and the constants A and k have to be included again in order to get R(p) as a solution.

$$R(p) = -Ak\left(-\tfrac{p}{k}e^{-kp} - \tfrac{1}{k^2}e^{-kp}\right) + A\left(-\tfrac{1}{k}e^{-kp}\right) + c$$
$$= Ape^{-kp} + \tfrac{A}{k}e^{-kp} - \tfrac{A}{k}e^{-kp} + c$$

$\therefore\ R(p) = Ape^{-kp} + c$ and as $q = Ae^{-kp}$ it has been verified that $R(p) = q \times p$

Assuming that the demand for a certain product is $q = 1000e^{-0.015p}$ [5] where A = 1000 and k= −0.015.

$$\frac{dq}{dp} = -15e^{-0.015p}$$

$$E(p) = \frac{-15pe^{-0.015p}}{1000e^{-0.015p}} = -0.015p$$

- Demand would be elastic when −0.015p > −1 $\therefore$ p < 66.67
- Demand would be inelastic when −0.015p < −1 $\therefore$ p > 66.67
- Demand would have unit elasticity when −0.015p = −1 $\therefore$ p= 66.67

=>Graph

$$R(p) = Ape^{-kp}$$
$$R(66.67) = (1000 \times 66.67)\, e^{(-0.015 \times 66.67)}$$
$$\therefore\ R(66.67) = 24525.30$$

$\therefore$ When selling the product to the optimum price of 66.67 €, for which a percentage increase in price would change demand in an approximately equal way, would result in a revenue of 2425.30 €.

Another example of a demand function is a quadratic one, $D(p) = q = (a - bp)^2$ [6] where *a* and *b* are parameters and *p* is the variable, so that *a* describes the autonomous level of demand while *b* describes how the demand *q* of a good changes when *p* changes

$$E(p) = \frac{p}{q}\frac{dq}{dp} \quad , \text{ so } \quad \frac{dq}{dp} = -2b\,(a - bp)$$

$$E(p) = \frac{-2bp\,(a - bp)}{(a - bp)^2} = \frac{-2bp}{a - bp} \qquad if\ \tfrac{a}{b} \neq p$$

Regarding the consequences on the revenue,

$$\frac{dR}{dp} = p\frac{dq}{dp} + q$$

$$= -2bp\,(a - bp) + (a - bp)^2 = -2bpa + 2b^2p^2 + a^2 + - 2bpa + b^2p^2$$
$$= a^2 - 4bpa + 3b^2p^2$$

Maximum revenue would occur when $\frac{dR}{dp} = 0$

5
http://www.ncssm.edu/courses/math/apcalcprojects/econ/Elasticity_of_Demand_Student_Handout.pdf
6 http://www.econclassroom.com/?p=2549

$$a^2 - 4bpa + 3b^2 p^2 = 0$$

Increase in revenue would occur when $\frac{dR}{dp} > 0$

$$a^2 - 4bpa + 3b^2 p^2 > 0$$

Decrease in revenue would occur when $\frac{dR}{dp} < 0$

$$a^2 - 4bpa + 3b^2 p^2 < 0$$

In order to find R(p), $\frac{dR}{dp}$ has to bee integrated.

$$\frac{dR}{dp} = a^2 - 4bpa + 3b^2 p^2$$

$$\int \frac{dR}{dp}\, dp = \int \left(a^2 - 4bpa + 3b^2 p^2\right) dp$$
$$= \int \left(a^2\right) dp - \int \left(4bpa\right) dp + \int \left(3b^2 p^2\right) dp$$
$$= \int \left(a^2\right) dp - 4ba \int \left(p\right) dp + 3b^2 \int \left(p^2\right) dp$$
$$= \left[a^2 p\right] - 4ba \left[\frac{p^2}{2}\right] + 3b^2 \left[\frac{p^3}{3}\right] + c$$
$$= a^2 p - 2bap^2 + b^2 p^3 + c$$
$$= p \left(a^2 - 2bap + b^2 p^2\right) + c$$
$$= p \left(a - bp\right)^2 + c$$

$\therefore R(p) = p \left(a - bp\right)^2$ and as $q = (a - bp)^2$ it has been verified that $R(p) = p \times q$

Assuming that the demand for a certain product is $q = (250 - 6p)^2$

$$\frac{dq}{dp} = -12\,(250 - 6p)$$

$$E(p) = \frac{-12p\,(250-6p)}{(250-6p)^2} = \frac{-12p}{250 - 6p}$$

•Demand would be elastic when $\frac{-12p}{250 - 6p} > -1$ ∴p < 13.89

•Demand would be inelastic when $\frac{-12p}{250 - 6p} < -1$ ∴ p > 13.89

•Demand would have unit elasticity when $\frac{-12p}{250 - 6p} = -1$ ∴ P = 13.89

$$R(p) = p \left(a - bp\right)^2$$
$$R(13.89) = 13.89\,(250 - (6 \times 13.89))^2$$
$$\therefore R(13.89) = 385802.47$$

$\therefore$ When selling the product to the optimum price of 13.89 €, for which a percentage increase in price would change demand in an approximately equal way, would result in a revenue of 385802.47€.

In order to give a precise example from my daily surrounding, I was intending to explore the Price Elasticity of Demand for Coffee in Germany for an average person. As coffee is said to be the most popular beverage of Germans, its consumption in general is constantly increasing or at least not expected to decline significantly in the future. Statistics suggest that the average German consumes about 150 liter of coffee per year, what accounts for approximately 1184 cups annually.[7]

In general coffee can be regarded as inelastic due to its addictive nature and the lack of substitute beverages with a similar effect. Moreover, certain brands of coffee with distinct flavors and a unique reputation such as Starbucks Coffee lead to a strong brand loyalty of its costumers, hence those coffee brands are largely inelastic so that an increase in price would not cause a significant decrease in demand. However replaceable types of coffee beans and powder, for example from the supermarket or bakeries, are rather of elastic demand because of the existence of substitute brands.

Another factor influencing the demand of coffee is the income of a person as well as its willingness to spend money for that certain type of coffee.

Assuming that a person consumes 1184 cups of coffee per year and has a relatively high income and a high brand loyalty for good quality coffee, it might be likely that an increase in price of coffee will only lead to a 10% decrease in demand.

By means of the demand function $D(p) = q = (a - bp)^2$

where the autonomous level of demand a is 1184, the decrease in demand per price increase b is 10% of 1184, hence 118.4, and p is the price

Therefore, $q = (1184 - 118.4p)^2$, so that $\dfrac{dq}{dp} = -236.8\,(1184 - 118.4p)$

$$E(p) = \frac{-236.8p\,(1184 - 118.4p)}{(1184 - 118.4)^2} = \frac{-236.8p}{1184 - 118.4p}$$

•Demand would be inelastic when $\dfrac{-236.8p}{1184 - 118.4p} < -1$

$\therefore p > 3.33$

---

[7] http://www.welt.de/print/die_welt/wissen/article10461591/Das-trinken-die-Deutschen.html

Hence, if this persons demand for that certain type of coffee is inelastic, a price increase for a coffee priced 3.33€ or above would not drastically change its coffee consumption and the person would continue to buy this certain type of coffee as the demand would not significantly change. As mentioned before this consumer behavior is observable in Germany for costumers who have a high brand loyalty towards unique brands such as Starbucks where a coffee beverage costs on average 3,50€. In case of a further percentage increase in price the demand for the Starbucks Coffee would thus not significantly decline.

•Demand would be elastic when $\dfrac{-236.8p}{1184-118.4p} > -1$

$$\therefore p < 3.33$$

Hence, This persons demand might be inelastic for other, more substitutable brands of coffee, therefore a price increase for a coffee priced less than 3.33€ would most likely change the coffee consumption and the person might consider to buy another brand of coffee as for lower priced coffee there is a wide range of substitute brands on offer. An elastic demand for coffee is especially observable for coffee to go from bakeries or cafes as well as coffee powder from supermarkets which are easily replaceable by another brand, hence an increase in price would encourage the consumer to change the coffee brand as he would receive an equal product for a lower price.

Regarding the implications on the revenue, maximum revenue would occur when $\dfrac{dR}{dp} = 0$

$$a^2 - 4abp + 3b^2 p^2 = 0$$

$$(1184)^2 - (4\times1184\times118.4)p - 3(118.4)^2 p^2 = 0$$

$$42055.68p^2 - 560742.4p + 1401856 = 0$$

Solving with the GDC, p is 10 or 3.33

As 10€ for a coffee is highly unlikely, 3.33€ is the optimum price for the consumer with a demand of 1184 cups of coffee annually and a demand decrease of 10% per percentage price increase, what shows that at 3.33€ the demand is of unit elasticity.

However, assuming that a person consumes 1184 cups of coffee per year and has a relatively low income and a low brand loyalty for good quality coffee, it might be likely that an increase in price of coffee will lead to 40% decrease in demand.

Therefore, $q = (1184 - 473.6p)^2$, so that $\dfrac{dq}{dp} = -947.2\,(1184 - 473.6p)$

$$E(p) = \frac{-947.2p(1184 - 473.6p)}{(1184 - 473.6p)^2} = \frac{-947.2p}{1184 - 473.6p}$$

•Demand would be inelastic when $\dfrac{-947.2p}{1184 - 473.6p} < -1$, $\therefore p > 0.83$

and inelastic when $\dfrac{-947.2p}{1184 - 473.6p} > -1$, $\therefore p < 0.83$

Hence, this persons demand is inelastic for coffee priced at least 0.83€ and elastic for coffee priced less than 0.83 €. As the price range of about 1€ for a cup of coffee is still considered average in normal bakeries or takeaways, the customer might remain loyal to this brand of coffee for convenience reasons or because he likes the certain brand of coffee even though it might not be of extraordinary quality. However its demand is elastic for other types of coffee, where a price increase for coffee priced less than 0.83€ would most likely change its coffee consumption and the person might consider to buy another brand that is even cheaper, for example by changing to lower quality products as he can't afford to continue buying the more expensive one.

Conclusion

In light of my investigation it has been shown how Price Elasticity of Demand can be helpful to predict a change of demand caused by an increase in price. By exploring two general cases of demand functions, the different levels of elasticity could be simulated and moreover related to the rate of change of revenue. Integration has shown that the concept gives reliable information about revenue of a good at a certain price as the formula of price Elasticity of demand can be further calculated and integrated to finally come back to the formula R(p)=q×p. By applying the concept to a real life situation, it became evident that also other factors such as income and brand loyalty affect the elasticity of demand and its resulting revenue.

Bibliography

"Elasticity of Demand", 11 november 2013,
http://www.ncssm.edu/courses/math/apcalcprojects/econ/Elasticity_of_Demand_Student_Ha
ndout.pdf

"Differentiation: Basic Concepts" , 11 november 2013, https://highered.mcgraw-
hill.com/sites/dl/free/0073051918/297329/ch02.pdf , p. 117

"Examples of Elasticity", 11 november 2013,
http://www.economicshelp.org/blog/7019/economics/examples-of-elasticity/

"Introduction to linear demand equations", 12 november 2013,
http://www.econclassroom.com/?p=2549

"Price Elasticity of Demand" , 11 november, http://www.tutor2u.net/economics/revision-
notes/as-markets-price-elasticity-of-demand.html

"The Relationship between Price Elasticity & Total Revenue", 12 november 2013,
http://smallbusiness.chron.com/relationship-between-price-elasticity-total-revenue-
24544.html

„Aktuelle Informationen und Statistiken zum Thema Kaffee", 24 november 2013,
http://de.statista.com/themen/171/kaffee/

"Das trinken die Deutschen", 24 november 2013,
http://www.welt.de/print/die_welt/wissen/article10461591/Das-trinken-die-Deutschen.html